THE APP COOKBOOK

THE EXPERIENCE OF
MAKING AN iPHONE APP FROM SCRATCH

STEP BY STEP PROCESS OF WHAT TO EXPECT WHEN CREATING YOUR OWN APP

By: Caroline Fielding

THANK YOU

This book is dedicated to Mom, Dad, Cathie, Dean, Bryan and Steven, and all my dear friends and family who always believed in me. You know who you are!

A LITTLE ABOUT ME AND HOW I STARTED ON THE LONG JOURNEY OF CREATING MY OWN APP BUSINESS

One spring day, I was in the process of boxing up some old items. I came across my high school yearbook. I opened it, cringing at the thought of seeing my "big" hair and bad makeup photograph. I then read some of the quotes from teachers and friends, and finally, came across my own entry which I had written immediately after my graduation, back in 1988. The entry was supposed to be about what your goals are in life, and where you see yourself in ten years. I wrote "I will be married, with three children, and be CEO of my own business. I will own a yellow Lamborghini and a huge mansion with money I made all by myself." My, what ambition I had even back then. I guess I always had the entrepreneurial "bug" in me. I had drawings of prototypes, pages and pages of ideas, marketing strategies for certain startup businesses, doodling pads of logos and names for companies I had dreamed up, and handwritten notepads all over my room. In fact, most of the little pieces of papers scattered in my school notebooks and in the bottom of my purse were ideas that came to me throughout the day.

None of these ideas ever came to fruition. I tried, but money, time, and life made it impossible. One idea actually made it through to a patent attorney who gave me a 95% chance of it being accepted by the U.S. Patent Office. At that time, however, my husband was laid off and my first son was about to be born. Life lessons, patience and perseverance kept the entrepreneurial fire under me, and I knew, *I just knew*, one day I would have that master idea that would be brought from my creative mind, into a complete product to share with the world.

I did, in fact, get married to my high school sweetheart, had two beautiful sons, and obtained my business degree after our second son was born. Unfortunately, the marriage ended in divorce after 13 years. I now had to share my two sons with their Dad and we did the best we could as divorced co-parents. I was working for someone else full time and, by the way, I was not driving a yellow Lamborghini, or living in a mansion. I lived in a 1044 square foot townhouse with my sons, and was very quickly bursting out of every closet, using every square inch of under-bed-storage space and storing my sons' lacrosse gear behind my kitchen island, out of sight from anyone who entered. The sure smell of teenage boy sporting equipment lurked, however. Thank goodness for those smell-good room sprays.

Fast forward to 2010 and I was not getting any younger. I wanted to work for myself so badly and bring all these ideas I had to life. I would need money and time. I was divorced, working full time, driving a 4 year old SUV, and raising two teenage sons. It was all up to me. I made a plan. I needed to become debt free, except for my mortgage, to be able to have some extra monthly income to put into any investment idea that may pop into my head. At that point, I had decades of notebooks and sketch pads, and while I believed in all of them, I did not think I had "the one great idea" yet. I knew it would come when I had everything in place. Funny how life works that way. It's like *life knows* the right time and place for certain things. My mother always said if you have to force something, it isn't right. If it falls into place without force, then it is meant to be.

I had a few little splurges in my life at that time. I had acrylic nails on, had biweekly pedicures, went to a salon to have my hair highlighted colored and styled, and I loved to shop. I also had movie channels with my cable company, a little credit card debt and a car payment. I wrote these items down on a legal pad. I am not the most techie person and prefer to write things out instead of using a software program or computer. I wrote my plan, and I gave myself one year to become debt free except for my mortgage. I called my cable company and stopped the movie channels, going only to basic cable. I reviewed my cell phone contract and cut that down drastically, and negotiated a better rate with my car insurance company. I called my credit card company and asked them to increase my limit, and picked a program with high reward points, so that when I was ready to start up my business, I would put most of my start-up costs on my credit card, and then get cash back with the reward points. I took off my acrylic nails, performed my own pedicures (I was not very good at first with red nail polish - someone asked me "what did you do to your feet?" - I guess it looked like my toes were bleeding), experimented with boxed drug store color for my hair (despite one bad "Elvira" color where I actually had to call in sick at work the next day to fix it, I finally found one that worked), took my lunch and breakfast every single day to work and basically cut out all extras. I found a thrift shop by my house that had half priced items every third Thursday. I would go there after work and I found a 50 cent laptop bag, 5 cent office supplies (I needed yet more notepads to sketch), a $2.50 winter jacket and things I could use to "craft up" and give as gifts, such as jars, baskets and bags. I went without all my "extras" for an entire year and I paid off my credit card, my car, and actually saved quite a bit of money and placed it right into my savings account. I told my loved ones around me that I was just on a budget and being frugal, and I was, but my plans for doing this were much bigger.

It was 2011 by this time and I was socking away the cash, and feeling very good about what I had accomplished. And I had found my "grand idea." I was going to own laundromats. I had the name for my laundromat picked out, I had researched laundromats in three different counties, joined two laundromat industry organizations and researched marketing ideas and strategies. I put a bid in on a laundromat 15 minutes from my home, knowing the owner was in dire need of getting out from under it. The owner did not accept my offer. What?!? I couldn't come up with any more money and he was not budging. It was very upsetting, as I had researched at least 50 different laundromats and had really thought I found a gem in a nice neighborhood, and had dreams of making it into a beautiful laundromat/social-type place with neighborhood customers I would know for years. I had it all planned out, knowing my customers and I would treat each other like family, just like the traditional old neighborhood laundromats where everyone knew each other (back in the day). However, as I mentioned above, my mother always told me if anything is forced, it is not right and not meant to be. The purchasing of this laundromat seemed forced. The buyer was almost resentful that he had to sell and did not make the experience very nice. The broker was a quirky, strange fellow who mumbled constantly and I could not have any constructive conversations with him. It is as if it was his first day on the job...every day!

As I was wallowing in my pity from yet another "fabulous idea" of mine not making it to completion, I drove to my sister's home about 30 minutes from my home. I had to get on the interstate. This is where my idea was born. A mini van cut me off, with an Asian woman driving. It almost caused me to have an accident and crash into the guardrail. Having an SUV, I could have rolled over across lanes and lanes of the interstate. I was so incredibly mad at this Asian lady driver who was oblivious to what just happened, and the light bulb just suddenly went off. I wished I had an old bus that I could just hit her with. I would never want to literally hit her, or hurt her. I was just so mad! The light bulb shone brightly! I was going to make a smart phone App where you could virtually hit someone with your own bus. I would make it fully customizable so you could hit anyone you want. Just like that! I went to my sister's house and had a lovely visit, with thoughts of my app idea swirling in my head.

That night, I got out my trusty notepad, put my phone on it and drew around it (remember how you drew around your hand during grade school to make a Thanksgiving turkey?) to get the right size, and proceeded to draw my vision of what this game would look like within the four corners of the screen. I have never been accused of being an artist. When I play Pictionary, my dogs look like cows and someone once blurted out "it's a spoon" and it was really a balloon. I wanted the characters to look exactly like who you want to hit. I am not prejudiced and have absolutely nothing against any race, ethnicity, nationality, religion, etc., it's just that I happen to have had an Asian woman cut me off and I wanted to hit that exact Asian woman. So I made all sorts of characters, from every ethnicity, race, etc. to choose from. This was going to be an equal opportunity bus. So, with my stick figures and bus drawings in hand, I set out to find a company to develop my app.

I immediately came up with a name for my app, and called it Bus Rage. I scoured the internet and app store, and researched the name and found out it had not been taken. I was getting excited. I searched and obtained my domain name, which also had not been taken. My mother's words were in my ear - if it just falls into place, it is meant to be. With that in my head and passion in my heart, I proceeded to go on the very long journey of creating my very first smart phone app.

A very tough, frustrating, exciting, ambitious year was ahead of me, and I loved every minute of it. I wanted to form a corporation so that I could create this app, and future apps, under the same company. My company, Dryven, Inc. is a combination of my identical twin sister's son's name (Dean) and my two sons' names (Bryan and Steven). It was so fitting because I have always been driven and the name is pronounced the same way. I immediately searched the Division of Corporations and again, this name had not been used and the domain name was available. Again, nothing forced - it all fell into place.

During this year of developing and designing my app, I continued on my budget, putting every spare red cent towards the creation which included the development, the marketing, social media, the website, domain names, hosting account, corporation paperwork, business checks and various other start up business costs. With my monthly bills very little, I had quite a bit to put towards the app, and even make some upgrades and changes that I may not have been able to do if it hadn't been for my proper planning. This year was tough financially because not only was I financing my business, but my oldest son obtained his driver's license and then you know what comes next. Car insurance. For a teenage boy. Then it was college applications, and senior year activities like homecoming, class ring, prom and senior pictures. My sons were good sports though, all they really asked for was pizza on Friday nights and they got it (I used a coupon of course). My point in saying all this is, if I can do it, anyone can do it. It takes discipline, perseverance, ambition, sleep deprivation (trying to work on the app after my full time job), and of course passion. I always believed in Bus Rage 110% from the very beginning and I literally saw it in my dreams at night.

My first app, Bus Rage, went live on the App Store on October 10, 2012. I was so excited when I got word, I immediately ran to my stairwell at work to calm myself down. I called my mother and father first (from the 29th floor in the inner walls of the stairwell of my high rise office building - that was fun - it was a "could you hear me now" moment) and told them that everything fell into place. They were so excited and they both said "I knew it from the beginning that this was the one!"

Bus Rage is a fun, hilarious app where you can run someone, anyone, over with your own bus. You are the driver, you put your own name as the driver's name, then you choose your target from a variety of characters from all ethnicities and races. Then you name your target, or victim. If you don't know their exact name, just describe them. My very first target on the game once it went live was "that Asian lady who cut me off." I hit her as a target on the first try and it felt so good! After all, she was my inspiration.

My second target was an ex-boyfriend, John, and my third was another description "the lady at my mortgage company" who had called me about a late fee and when I told her I've never been late on a payment, ever, she suddenly saw she had me mixed up with someone else! I decided to Bus Rage her! It was instant therapy. This app is designed to be fun, engaging, and can be used by anyone, whether they are 7 or 70. Who doesn't want to hit someone with a bus on a daily basis?!?

Bus Rage is one idea that finally made it to completion. Now that I have one app under my belt, I am in the process of creating my second app, and hope to launch that app later this year.

My second app is another game app. Each player has their own shopping cart and the player strives to catch as many items in their cart as they can, without toppling it over. This is harder than you think and very addicting!

Having been through the ups and downs of creating my own app from scratch, and basically treading water throughout the entire process, I wanted to share my experience with others to help them along the difficult, but rewarding, road of app creation. I sincerely hope my experience will help others who have a fabulous idea for an app in their head, but do not know where to start. If I can do this, anyone can do this!

Start with this simple exercise: chin up, shoulders down, and smile on! Now...read this and then go create your own app!

Caroline

SEND THE INVITATION

First and foremost, before we dive into what to expect when creating your app, be smart about your creation. You have the idea to throw a dinner party, but you do not want to let the word out just yet. Keep quiet about it and do not go around telling everyone about your grand idea for an app just yet either. You want to protect your idea, your baby. Trust me, after spending 18 months creating my first app, it really *was* like my baby! So, protect your idea, keep it in your thoughts and your thoughts only, for now. Later down the road, you can share your exciting news, but only when the time is right. If you already have a name for your app, keep that to yourself also. The name of the app, the concept, your planned social media site names, your expected company name - really anything that you have thought of - can be stolen, so be very careful about revealing your ideas to anyone.

I kept Bus Rage all to myself for an entire year. I only told my Mom and Dad, and my development company that I hired to develop the app. That's it. Not my best friend, not my children, not even my sister. I had a certain reason for not telling these people, other than protecting my idea. I was not worried in the least about those particular people in my life stealing anything from me. The reason was almost 100% ego. I thought that if I succeeded, it was all mine, and mine alone. I could say I did it without any help from anyone and I wanted to feel this sense of accomplishment. I have always been like this, and sometimes it comes back to haunt me and other times it does not. However, this is just how I am wired, and how I feel most comfortable in handling my ventures. Whomever you decide to tell, do it with caution and advise them of the confidential nature of what you are telling them. It is really up to you to decide who to tell. Sometimes you need moral support, business support, financial support, or perhaps want some constructive feedback on levels of the app, pricing, or perhaps something else. However, if you do decide to tell someone who you do not really trust 100%, obtain a Non-Disclosure Agreement prior to telling them. You can download a Non-Disclosure Agreement from any legal form site on the internet. Have the company or person sign it first, before you tell them anything.

You may think this is overboard, but trust me, when your app sells 150,000 downloads and you suddenly have some cash, it is amazing who will turn on you. Just be careful.

When I thought of how to tell readers what to expect when creating an app, I thought it would be fun to create analogies with my other favorite passions – cooking and entertaining!

When other people asked me how I created my app, Bus Rage, and how to go about creating their own app, I found it hard to just dive right in and tell them. I didn't really know where to start. The idea came to me that it was just like planning a dinner party. Picture this scenario – you invite friends and family to a dinner party using a beautiful invitation. On it, you write down the place, the time, and the date. Then, for weeks prior to the dinner party, you plan your table setting, decorations, lighting, music, and so forth, and you have a sense of excitement about the planning phase of the party. Regardless of your excitement, you do want the details to be a great surprise for your guests, so you keep all the visionary details to yourself until the grand reveal. When creating your app, simply telling your close acquaintances and friends you are working on a "personal project." This is really all you need to reveal at first. Then, a time will come down the road to tell everyone, scream from the rooftops, and throw a launch party for your app. Until then, just "send the invitation."

PREPARING TO COOK

It is Thursday. You are hungry. Starving, if truth be told. You wonder what to prepare for your dinner party which you have planned for Saturday night. Sometimes, you know exactly what you want to make, much like I knew exactly what I wanted my app to look like and how it would work. Other times, you do not have the slightest idea of what to cook, what to serve, and need inspiration.

So, if you already have an idea for an app, great. That is a huge first step. Those who do not have an idea for an app yet, but just want to get into the app creation business and find some inspiration, can go to the App Store and browse the categories. This is largely similar to looking through a cookbook to find your perfect recipe. You may know for sure you want a chicken dish and browse the recipes surrounding chicken, or you may not know what you want to make at all. Browse through the cookbook and you will surely find a dish that makes your mouth water. You may also see a dish that looks pretty good, but you decide to modify the dish to suit your individual taste. Sometimes a recipe you come across, will put you into an entirely different direction from when you first starting thumbing through the cookbook. You may have been looking for a meat dish, but you come across a Spanish Bean and Rice dish so succulent, so mouthwatering, you just have to make it. So, if you do not have a specific idea yet, browse the categories in the App Store for inspiration. Surely, something will pop out at you and you will come up with a fabulous idea, as you are making preparations for your dinner party menu.

PICKING YOUR CUISINE

Picking your cuisine is much like planning a theme for your dinner party. You may want a Mexican Fiesta theme with enchiladas, queso and chips, and seafood tacos with fresh guacamole. Or you may want to go another direction altogether and prepare barbecue such as grilled hamburgers and hotdogs, spicy french fries and chili on the side. Much like your food would fall into certain categories or themes within a cookbook, your app must fall within certain categories within the App Store.

The following are categories on the App Store that your app will need to fall in to:

- Books
- Business
- Catalogs
- Education
- Entertainment
- Finance
- Food & Drink
- Games
- Health & Fitness
- Lifestyle
- Medical
- Music
- Navigation
- News
- Newsstand
- Photo & Video
- Productivity
- Reference
- Social Networking
- Sports
- Travel
- Utilities

- Weather

SETTING THE TABLE

Once you have mentally prepared yourself to cook, chosen your recipe, ingredients and cuisine, you are ready to begin. It is the night before the dinner party and you are about to set your beautiful dinner party table. Picture your table. Picture your app.

Will it be a checkerboard tablecloth and paper plates for barbecue, or elegant orange plates for fancy enchiladas? Where will you place the ice bucket? Will you use low lighting or colorful paper lanterns? Visualize your table, just as you will visualize what your app will look like. What tone are you planning to go with? What colors will you use? Will it be "elegant," "casual," "practical," or "funky." These are all details to pay attention to, in order to create your dream app.

APPETIZERS

Okay, so now you have your idea and the concept in your head. Now what happens? This is where most people get stuck. They just do not know where to go from idea to completion. To move forward, I recommend performing research, research, and more research, at this stage. Everyone serves cheese and crackers, or a vegetable platter. While there is nothing wrong with that, it has already been done. You have just come up with an even better idea to serve your crackers, such as top them with cucumber, almond slices and bleu cheese crumbles. Fancy, huh? Has anyone ever served that at their dinner party in the neighborhood? You need to find out so you are not copying another idea that has already been served. You do not want to waste your time and money on an exact idea that is on the App Store already. Search the App Store thoroughly, Google your idea, use other search engines, look at app review sites, app blogs, app developer sites for their portfolio sections, and social media sites. Search high and low for your idea. Think of what key words you would use for your own app, and search those key words. Scour the internet, go deep within the pages of Google, and perform your due diligence. For example, I knew the name of my app was going to be Bus Rage, but if you do not know the name of your app yet, at least you know the concept of the game at this point, and what category the app will fall in to. Use key words that describe your app. I also knew the flow and concept of the game and how it was going to be a game where an angry person, or enraged person, could virtually run over someone else with their virtual bus. This is what everyone wanted to do in real life, at least in my opinion. So, taking Bus Rage as an example, I was performing my due diligence late one night, and I searched for the following key words:

- Bus Rage
- Rage games
- Driving rage games
- Driving games with bus
- Driving a bus app

- Apps with buses
- Apps with rage
- Rage app
- Angry bus app
- Angry rage with bus
- Angry bus rage
- Mad angry games
- Angry game
- Mad game
- Mad at someone app game
- Enrage games
- App store rage games
- Games containing bus
- Running over another person game
- Virtual bus game
- Virtual rage game
- Virtual angry game
- Customizable bus games
- Customizable rage games
- Custom bus rage
- Custom driving bus game

You get the idea. Be creative and spend days searching the App Store and internet sites to make absolutely sure your exact idea has not already been created. This brings me to something I would like to share with you. Of course, you want to think of an idea for an app that is so incredibly unique and fresh, that the entire world will surely jump at the chance to download your app. However, sometimes you do not have to re-invent the wheel. The idea is already out there, you just need to find a way to put another twist on it, make it even better, or modify it to suit another requirement. Just make sure you do make the app significantly different or modified enough to make a difference to users. The App Store does not like too many of the same types of apps, and you run the risk of being rejected if your idea is too much like another app that is already out on the App Store. If you find your app idea is already in existence, make your modification very unique.

Think about, and picture, a grilled cheese sandwich. A traditional grilled cheese sandwich is made with butter, bread and cheese. Sure, you can make it a "four cheese" grilled cheese sandwich, or use rye bread instead of traditional white. However, you need to be even more unique than that. Create your own grilled cheese sandwich with bacon, tomato, pickles, ham, eggs, and so forth. So, just because an app idea is already out there, do not let that deter you. Let it be inspiration to modify that idea to meet another requirement, or make it even better than the app already on the App Store. Think outside the box and modify it. Like the appetizer scenario above, if you found out your neighbor down the street served crackers, cucumber, bleu cheese crumbles and almond slices at her last New Year's Eve party, then modify the ingredients. Maybe cut up french bread, place cubed cheddar cheese and cubed cucumber on top, and then sprinkle it with sesame seeds. Both are delicious appetizers, yet are different enough to make each enjoyable and nobody will accuse you of being a neighborhood copycat. You want your idea to be so unique and marvelous, and especially want that idea to belong to nobody else but you. If you do not find your app out there, this is a great sign. Now you have silenced the hunger pangs in your stomach just a tiny bit at this point. You are still very hungry, and have a long way to go before the main course.

At this stage, you have your app idea and it fits into one of the above App Store categories.

In parallel, you are finished browsing through your cookbook, breezed through your appetizers and picked out your main dish meal that you will serve to your gracious guests. No other person in the neighborhood has ever served this meal before. You are going to throw the best dinner party on the block. Now, you will have to make sure you have all the time, ingredients, equipment and space to create the perfect main meal and serve it to your family and friends.

Look around your kitchen to see if you can create the meal with ingredients in your freezer, refrigerator and pantry. Look around your personal life, to see if you have the "ingredients" and "equipment" to complete your app.

You will need money, a budget, patience, time, vision, creativity and the willingness to see it through until the very end. To some, money will be the easiest item to get, others may find it hard to be creative or envision what the app will look like and find it hard to express their ideas to someone else.

Budget is huge of course. Unless you are a huge corporation, have family money, are independently wealthy, or are just plain lucky to have quite a bit of spare change hanging around, you will need to figure out a budget to pay for the development of the app. If I can do it, so can you though! You can stay on budget and create a financial plan in order to pay for your dream to come true. Trust me, if you want to find a way, you *will find a way*. Make a 6 month plan, a 12 month plan – you know the old saying "time will go by no matter what," so might as well make the most of it.

Patience, keeping your cool, being the consummate professional and staying positive is very important. You are building an app, you are now your own entrepreneur. Act like one and always remember to treat this project like it really is – which is your own business. You will need time and a desire to go through months, perhaps years, of working alongside others to make your app come to life.

You also must have creativity and vision. You do not have to be a good artist, but you do need to be precise and describe what you want to the developers and individuals assisting you in the creation of the app. A very clear picture and the ability to relay information in a concise, brief manner is also very helpful.

I have spoken to other customers and app developers around the globe, and this is a pretty good idea of what you will need as far as funds, and the range of what you can expect each item to cost:

Complete App Development: $5,000 - $75,000
Website Design $300 - $1,000
Corporation Documents $100-$200
Domain Name $10-$20 per year
Hosting Account for Website $50-$300 per year
Social Media Site Design $1,000 - $4,500
Logo Design $300 - $1,500
Marketing $100 - $5,000

Now let me break this down a little more. If you taught yourself how to write the code, learn Apple's criteria and requirements and can create and design your own app, which is fantastic. You will have to put in hours upon hours, but you will save a heck of a lot of money. Most of the money you will spend will be put towards your development costs. If you decide to teach yourself how to write the code for your app, you do run the risk of not complying with Apple's rules. If you are going this route, definitely do your homework and read and then re-read the Apple Developer Manual which can be found online. I know of an app developer who made a photography app, worked two years on it in his living room teaching himself, only to be rejected by Apple three times, before he finally was accepted. This took another year, making his app development three years in the making. By that time, the App Store was saturated with photography apps.

If you hire an app developer, like I did, research the company thoroughly. I will teach you how, and talk about the steps to take when hiring a developer in detail later in the book.

Much of the above budget list is negotiable and you can do all, or just some. If you are on a strict budget, really all you need to do is pay for the App Development. The other stuff is really just "fluff" and is not one hundred percent necessary, at least in my opinion. If your one and only goal is to get an app for sale on the App Store, then getting that app created and developed is all you need to do. The rest can be added later. There is nothing saying you need to do this all at once either. Once you pay down the developer costs, you can have a website created. Then three months later, create a Facebook page, and so on and so forth.

I kind of fell into the "I better do it all at once so everything comes out at the same time and Bus Rage can be a full, 100% completed package at launch time." In retrospect, I think this was a mistake and I think some of the items above could have waited. I think the website, the Facebook Fanpage and domain names could have waited until I made some of my money back and I really didn't have to strap myself financially in the beginning trying to get all of the above out at the same time. Live and learn, and I am hopeful that my experiences assist you in your development phase. With my second app, I am creating the entire package in smaller steps, instead of all at one time. This is less overwhelming, and I am not as strapped every single month until the app launches.

SALAD

At this point, you have your idea for your app. You have researched the concept. The app fits into a category on the App Store. You know you have the passion to see it through and know you can make the finances work in order to create the app. Now it is time to name it.

If you do not have the name for your app, now is the time to name your app and bring it to life. Fun, catchy names for games work well. If it is financial app, or a productivity app, describing what the app will do in the title helps the user search easier for the service the app is intended to provide. If, for instance, you are creating an app to teach people how to play the piano, a name like Pianocchio would be fun, and the user would know the app has something to do with music and the piano, just by the name. If you have trouble naming your app, please email me and I will try my best to come up with a great name for you, free of charge. I will list my contact information and email address at the end of this book. One of my creative passions is to name products, apps, and books, and I love to help other app developers.

Once you have decided on the name and it is available, it is time to write everything out. If you prefer to type it out, or write it out, it is really up to you and is strictly personal preference. However, you do need to document the app, the name, and the date of the writing, for proof that you were the author and creator of this app.

Think about your audience that you are trying to reach with your app. If you are trying to reach the teenage crowd, the baby-boomer crowd, the Wall street types, or the stay at home moms, keep this in mind throughout the creation of your app plan. Write out your app plan like you are one of your users. Put yourself in their shoes and think about what you would want to see in this app, and begin your outline, or App Plan, as I like to call it. This App Plan will be given to potential developers so they have an idea of what your app is all about. This is an entire overview, an entire description of every detail of your app, so they have a clear, concise idea of your vision.

My App Plan had the following headings:

- NAME OF THE APP
- CREATOR OF THE APP
- DATE OF CREATION
- CONCEPT
- SPLASH SCREEN
- CONCEPT OF LEVELS/MENU
- WHAT THE APP WILL LOOK LIKE
- BACKGROUND OF THE APP
- MUSIC TO PLAY
- DESCRIPTION OF LEVELS
- WHO THIS GAME IS MADE FOR
- POINT SYSTEM/SCORES
- SOCIAL MEDIA INTEGRATION

For my first and second apps, I used this same outline.

At the top, place the name of your app, your name (or your company's name), and the date. On the next page, start with Concept.

CONCEPT

The Concept is basically a one paragraph description of the game. If you looked at a cookbook and you saw "Chicken Bruschetta" you would know basically what this dish would look and taste like, even though you may not know the exact ingredients or measurements yet. Think of this Concept paragraph as an overall view of the app. "What's the Plan" is another way to think of it. Using our example above, about the app that teaches you how to play the piano, you would say in the CONCEPT paragraph something like this:

> Pianocchio is an app which will teach you how to play piano even if you have no prior experience on the piano keys. Step by step instructions will teach you in 30 days how to play 3-5 basic songs that everyone knows and can sing along to. Once you master a song, you have the option to share this on your social media sites. This app will have a timing option, to give the user more of a challenge.

It is just that simple. Do not overthink this. If you are sitting at a restaurant and someone asks you "What does your app do?" This is a short paragraph explaining exactly what it would do. It does not have to go into intricate detail at this point. That will come later.

SPLASH SCREEN

Second, you need to design your splash screen and logo. Picture your smart phone and the little icons that show what apps you have. Some are colorful, some are plain, some contain the name, and some just a logo. You need to design your logo to make it stand out. I did not draw my logo, but I did design it. I wrote out a description of exactly what I wanted it to look like. I cannot draw for the life of me, but I can envision what I want something to look like, and I can describe it well. So, just write down a description and you can hire someone to actually make your logo and splash screen. The splash screen is what happens when you click or hit the app icon. Sometimes it is a short, animated video and sometimes it is just a splash screen before the menu pops up.

So, for the logo and splash screen, in using the example of the piano app, you can write something like this:

> Pianocchio will have a black piano with 2 long candlesticks containing two red candles which are lit and sit on top of the piano. A bouquet of gold plated flowers in a blue vase will be right beside the candles. A book containing music lines will be open, and ready to read, underneath the lit candles. Once you hit the icon, it will be a short, 5 second video that will pop up with beautiful classical music playing, the keys of the piano moving up and down and a "splash" that says "Pianocchio Will Teach You How to Play this beautiful music!" Then, the screen will switch to menu options. Basically, think of something that will engage the user and make a fabulous first impression.

Picture a cooking show and how they entice you by showing you clips of what they are going to make. A clip of a pile of chocolate chips, a screen shot of strawberries and funky music playing in the background will certainly entice you to keep watching and "enter" the cooking show. What will entice and engage your user? This is the premise of the splash screen, because it is the first thing the user will see when they open your app.

CONCEPT OF LEVELS/MENU

Next, write out how the home screen will look, which is the menu. For a game app, you would want to put "How to Play" or "Description of the Game," "Start," "Resume Play," "Settings," and "Pause."

Think of using this app for the first time. What would you want to see on the home screen and what would you want to know regarding this game or how to play this game. This is what you want to put in this section of your plan.

Write out a complete description of the game and how to play. Explain how to gain points, how to post to social media sites, how to navigate within the game and how to pause/start.

Now you need to write out the levels, or milestones you want to have in your app. For instance, for a driving game, you may choose to launch 10 levels at first, then update the game every two or three months with 5 additional levels each update. For the Pianocchio app, you may want to make each level a milestone such as when you master a Beethoven set, or make it a timed level where you need to play certain keys in a certain number of seconds to move on to the next section.

Basically, what each level will look like, what you do on each level, and how you get to the next level. For a productivity app, how does the user get to certain details, how do they go back, and how do they go forward. "Pause" and "resume" buttons are important because people are busy or need to do something else for ten minutes, and want to come back to the same place they left off.

WHAT THE APP WILL LOOK LIKE

Be very descriptive of the items in your game. For instance, make the piano blue, or make the keys to the piano red. If you are making a game, describe the characters, or bus, or car, or aliens as far as color, shape, whether you want them more animated-like or more stick-figure-like, or if you want it to look like a real life car, or bus, or alien. If there is such a thing as a real live alien. My mother would say there is, but that is a discussion for another book. I digress. To put this into foodie terms, this is where you would describe your food to a cookbook photographer or creator. Picture telling him that you want your chocolate brownie cake on a yellow plate, with a strawberry scoop of ice cream on the side with a sprig of mint sticking out of the ice cream, with chocolate drizzle around the sides of the yellow plate. Your description of the items in the app matters, and this is a huge item to pay attention to when outlining your app plan.

Do not hold back and it is better to be overly descriptive than leave out details.

BACKGROUND OF THE APP

The background is also important. Just like the chocolate brownie cake with the yellow plate is a huge part of the photograph, putting the plate in front of beautiful, picturesque window with mountain countryside in its view is even more spectacular. Place special attention to your background and describe this in your app plan.

Think about the background of each level, colors, images, pop ups, and font size are all very important aspects to pay attention to.

MUSIC

Something to consider, which honestly I did not even think about during the initial phases of my first app, is music. Just like a beautiful, romantic dinner and dessert is complimented by romantic music, or perhaps old school rock and roll music, an app sometimes is complimented by music as well. A fun, upbeat song for a physics game, or the soft, spa-like hum to go along with an app depicting travel destinations, or a blaring rap song for a graffiti spraying app goes a long way in complimenting the entire playing experience. It is just something extra special to give to your users, and you need to decide now if you are going to use music or not.

DESCRIPTION OF LEVELS

I wrote the description of my game myself, although if you are not a clear, concise writer, you may want to get a trusted friend or family member to do this. This description will be on the App Store or Android and they will be the first words any potential user will see. These words will make them decide whether or not to buy your app. So make this count! If you were in front of a panel of 5 food critics, and you had to describe your awesome beef stew that you just made, how would you describe it? This is what you need to keep in mind. How would you describe your app to someone else? Engage the reader, show how the app will make his or her life better, how fun it will be to play the app, how it would help them in everyday life, how it would improve their productivity, or how the app would help them find their local events. Whatever your app does, describe it here. Be positive! The key is to keep it brief; however, because nobody likes to read a "book" when deciding whether to buy your app. Give them a detailed taste of the app, without going into total detail. Just tell them the basic details, but save the surprises for when the players play the actual game, or use your app the way it is supposed to be used. To use the beef stew as an example, you are engaging the food critic panel with your ingredients and telling them you used beef stew that will break apart with their fork, and carrots that will melt on their tongues, but not necessarily telling them that once they bite into that carrot, they will taste some brown sugar and basil leaves.

If you are going to have a point system in place for your app, this is the time to write this down also. Whether it is gathering certain items for a certain number of points, or when you finish a level in a certain amount of time, you obtain points, and if you finish sooner or later than the last time, you gain and/or lose points according to the time, it is important to write this all out and lay it out in black and white terms. Sometimes this is hard. I can equate this to a cooking class I took years ago. It was a fun class, and the teachers gave the students points for taste, texture, presentation, timeliness and cleanliness. I did not do so well my first time, so I went back twice more, to try to boost my total points. The class definitely left me hungry, no pun intended, for more. This is what you want to do for your players or users of your app. Make it a challenge and fun, but leave room for them to want to come back and better their score over and over again.

WHO THIS GAME IS MADE FOR

If you are making an animated fish game, it is likely you are targeting younger gamers. These days, there are kids on tablets and smart phones as young as four years old! I am in awe of the children nowadays with the way they adapt to technology and immediately know how to work games and products. I remember growing up when all we had was Atari and "trouble shooting" was taking the cartridge out of the Atari console, blowing on it hard to get the dust off, and sticking it back in to see if the game worked. And you know, nine times out of ten, it worked! Now, there are about 15 steps you can take to troubleshoot for any one game. But I digress. You must know who you are targeting when creating your app. A productivity app will most likely be aimed at 25-45 year old executives so you want the app to be polished and fairly easy to navigate, as busy, young professionals do not have a significant amount of time to master a new app. Children have a lot more patience, so perhaps add a few more levels, more colors and extra bonus items. A weather app is geared towards travelers and sports fans so keep the app entertaining, yet simple and easy, so they may quickly push a button to see if rain is in the forecast for their mountain biking weekend, or football game.

Know who your audience is, and put yourself in their shoes. Think of what, as a user, you would want to see, feel, hear and touch, then gear the details of your app towards those senses.

SOCIAL MEDIA INTEGRATION

Will the user be able to view the top scores of other users? Do you want players or users of your app to share the information, comment on the app, or share their score on social media sites? This is something you should decide now. This has the potential to make your app go viral. Just think if I shared the activity of Pianocchio to my Facebook account and it said "Caroline just scored 6000 points playing Pianocchio," then 12 of my friends commented on it and 28 "liked" it. Think of how many other friends on their sites would see this and the app will quickly snowball. If you do this, however, remember that you will need to pay for the Facebook Fanpage and Facebook integration within your app, so if this is something you want to do from the beginning, meaning you will need to pay for the creation of your Facebook Fanpage before the app is launched, as opposed to waiting and adding a Facebook Fanpage down the road five months.

MAIN COURSE

You have finished your appetizer and salad. Your dinner guests are now up out of their seats and mingling in the backyard deck with a cocktail, almost ready for the main course. The conversation is headed into fun discussions and European vacations and everyone is excited and upbeat. You have compiled your ingredients, arranged your kitchen equipment, chosen your main dish and side items, and you wrote out the itinerary for the evening. Now it's time to execute the meal, put it all together, place it in the oven and make your dish come together into a beautiful creation.

Just like some people hire a chef for the evening, I chose to hire a developer to develop my app. I am not a techie person, I am more of an "idea" person. I can envision ideas, but sometimes find it very hard to make them actually appear in real life. This "main course" involved hiring a developer. Remember what I mentioned above? The very first thing to do is to have whomever you tell about your app, sign a Non-Disclosure Agreement. This is extremely important because without it, anyone can steal your idea. With it, if they do steal your idea, you are entitled to damages and have legal grounds to stop them from going forward with the same concept and designs as your app. So, remember, first and foremost, before you talk to anyone, have them sign an NDA. After the NDA was signed, I sent each company my App Plan.

When researching a developer, the first thing I did was go to the internet. I tried a few local area developers that claimed to develop mobile apps, a few out-of-state developers, and a few out-of-country developers. I also looked around on internet blog sites to find out who developed mobile game apps. I narrowed it down to about five companies. From there, I asked for prior client references, performed due diligence by looking up the company, checking to see if it was a legitimate corporation listed in the State's Division of Corporation public records online systems, checking the Better Business Bureau for any complaints, asking for references and calling prior clients to discuss their experience with the company, Googling reviews on the company (some came up with bad reviews). Communication was key also. If a company did not timely return my phone calls or emails, they were put lower down on the list. One company was very persistent, even pushy, but I thought this was better than giving me the silent treatment. They seemed very professional and were willing to sign the NDA right off the bat, and enter into a professional, multi-page contract. If you would like the name of the company, please contact me and I will be happy to give you the name and comment on the overall experience. The development team was professional, they never asked for any more money, and always kept in constant communication with me. However, there were setbacks which I will talk about later; but, the overall experience was positive and in fact, I am having them work on my second app right now.

Consider the overall theme of this book where it ties cooking and mealtime with the creation of an app from scratch. Some chefs are so quick, they can literally whip up a dish in half hour by tossing all the ingredients in the air, so to speak, and grabbing them in a large bowl and sticking it in the oven. Other cooks whip up the same dish in one hour, cleaning as they go along and basically putting all the ingredients together in a more methodical fashion. App developers are the same way. Some are gung-ho, stay on time, charge a little more for meeting early milestones, and are not quick to answer emails or phone calls. This does not mean they are a bad company. It is just their corporate culture may be different than the company who is slow, communicates well, but may not necessarily meet all the deadlines. Many development companies are reputable and will do a good job, but I just want you to be prepared for the differences in development styles of app developers.

After you find and decide on your app developer and you are happy with them, it is time to sign the contract. A few things about the contract you sign with your app developer that I would like to share. First, all monies should not be paid up front. There should be milestones that are outlined in the contract with a percentage of money to be paid at the completion of each milestone. For instance:

Initiation of the Project/Initial Designs 30% of the total amount

Wireframes/Scope of Work 20% of the total amount

Delivery of the App/Upload to App Store or Android 30% of the total amount

Completion of the App and delivery of the source code 20% of the total amount

So, if the total amount of the entire development of the app is, say, $35,000, then you would pay:

Initiation of the Project/Initial Designs $10,500.00

Wireframes/Scope of Work $7,000.00

Delivery of the App/Upload to App Store or Android $10,500.00

Completion of the App and delivery of the source code $7,000.00

This way, you each hold a little bit of the power. If you pay them all up front, then they can take as long as they want. If you hold off on the entire payment until you are 100% satisfied with each milestone, the company may rush through the process. This way, everyone is more apt to stay on track. Do not contract with any company who will charge the entire funds up front.

Sometimes there are changes to the Contracts. Do not be afraid to ask questions and modify the Contract if you feel it is in your best interest. I have made "Appendixes" to my contracts to include certain items that I felt were necessary to be included in the contract. I like to be extra specific so there are no ambiguities down the road.

Just like hiring a catering company for your wedding or company picnic, you only give a deposit to hold the catering company to a date, then pay another portion for food purchase, then the remaining balance at the end of the event when everything is cleaned up and no guests have gone to the emergency room for food poisoning.

Once the contract is in place, you have milestones to look forward to and the designing of the app begins. This is a long, tedious process with multiple communications sometimes on a daily basis, for feedback, modification of ideas, and status checks.

This "Main Course" chapter is a big one, because there is so much to cover. This is where, during the course of creating and developing my first app, setback occurred, timelines were re-invented and the app took on a life of its own.

Keep in mind this will not be a smooth process. Throwing a dinner party for 50 guests is no easy task. You start days in advance, weeks actually. You make lists of groceries, caterers, table centerpieces, decorations, candles, playlists for music and choose dinnerware. Even with careful and meticulous planning, things go wrong and plans change. Keep your chin up, shoulders down and your smile on, during these times. Just think of your very own app on the App Store or Android one day and being able to tell all your friends and family that YOU created this app from scratch!

Cookies for everyone! Chocolate chip, lemon crumblers, peanut butter, and oatmeal are favorites. No matter how many times you have made your grandmother's chocolate chip cookie recipe, there will be that time that you burn the batch. It just happens. We have all wanted to try a recipe for the first time and when it came out of the oven, it was plain awful! You end up ordering a pizza! What do you do in these situations? You throw out the batch of cookies or main dish that you made, put back on your apron and try it again, this time adding a little more milk, using less butter, or incorporating more herbs and spices to make a dish more flavorful. This is not much different from creating an app. Sometimes you have to scrap what you thought was going to be so good in the beginning, and modify it to fit what can be done on a small screen. The visions for Bus Rage that I had in mind did not work correctly on a small iPhone screen. The integration of certain multi-media items, levels, or animations did not work out very well and I had to modify my game several times with the creative team.

There were also items that I visualized that were logistically too complicated and would take years and years of technology with no guarantee of it ever working. I had to constantly tweak the designs and levels and make changes to make it feasible. Some of these changes really made me upset and definitely tested my patience during the creation process. In the end, being my very first app, it was probably not 100% of what I had visualized in my head when I first started out, but it was pretty darn close and I am proud of the end result. I did learn a lot though, and my second app is going much smoother. If you keep making a tilapia recipe over and over, you will eventually perfect it and the cooking process will go much smoother and you will feel much more confident of your end result. It is the same thing with your app business.

Be prepared for time-line set-backs. Just like dinner is scheduled to be out of the oven by 6 p.m., you cut into the meat and it is still pink, so another 15 minutes in the oven is necessary. Then a dinner guest is running late so you put the dish back in the oven on low to keep it warm. Keep checking on it, nurture it with liquid, and cover your side dishes with tin foil to keep in the heat. Do everything and anything and do not give up. Your guests will finally show and then the evening can progress into a romantic, beautiful night. What seemed to get started off on the wrong foot, or head down a different path than intended, does not necessarily mean the end of the night will be lost. Never give up and always remember to light those candles. Dessert is coming, but you need to get through your main course first.

During this phase, your developer will be in close contact with you. Descriptions and modifications will be made mostly by email. Do not hesitate to change colors, sizes, or the overall look of the items on the screen. Make sure all items are to scale and look the way you want them to look. This may take a few back-and-forths to get this right, but in the end, think of how the user or player will feel when looking at this for the first time. Again, this will be mainly accomplished by email and conference calls. A few faxes or scans of drawings or modifications can also be expected at this phase.

The splash screen will be created and designed, the logo for your game, music to match your app will be sent to you for your review and feedback. It is very important for you, as the app creator, to give timely feedback. If you do not timely give your feedback, days or weeks can go by and nothing will be accomplished. You do not want this project at any time to come to a standstill. You want this to flow and become an ongoing, flowing river of ideas. Just like you would not leave your salmon on the grill for two hours, you want to keep a close eye on your app development progression and make it your priority. Remember, you are making this from scratch and this requires careful attention to many, many details over the course of many months, perhaps even a year.

Just like a recipe sometimes needs modification and substitutions, the development of an app is the same way at times. Just like milk can be substituted for egg, sometimes you have to modify or substitute something in your app that either cannot be done, is too complicated to be done or is too costly. Patience, the ability to be flexible and the art of compromise is crucial when working with your developer on your app. Picture yourself as a sous chef who works closely with a head chef at an exclusive restaurant. You have a fabulous idea for a new lamb dish. When you, as the sous chef, works hours on perfecting the dish, sometimes the ingredients used are just too costly to justify and pass along to the customer, so the head chef suggests modifying the recipe with less costly ingredients. Sure, the original idea was superb and you wish you could serve that wonderful, though costly dish to the patrons; however, the first night you did serve the modified version of the lamb, it sold out. So, you see, compromise and patience and putting your ego aside are essential in creating your app. Now, I understand this is hard to do. I am guilty of thinking of my app as "my baby" and only I know the vision I have in my head of what it will look like, how it will flow, what the player will feel, how I want the music to play, and so on. But, I had to make major compromises with my first app. There were many setbacks, delays and changes and although I did get upset multiple times, I discovered that this is just the nature of the business, and during the development of my second app I am much more prepared for the unpredictability of each stage of the development process.

Now I am going to talk about setbacks. In the kitchen, it is fun to cook with wine. I've done it a few times with white wine which is wonderful in a linguine. I was cooking for a boyfriend one night and I used a little too much wine and the skillet shot up in flames. In flames! I quickly put it out but the smell of burnt pasta was throughout the entire dish. It was gross. I threw the entire batch of linguini out and we had grilled cheese sandwiches and frozen french fries instead. It was upsetting but in the end, it was not the end of the world. I will now tell you about the multiple setbacks I encountered during my app development.

My development company had a fire in the backup server room of their facility and lost 45% of my data weeks prior to the completion date of my app. That was the worst and I literally cried over this. It was like cooking Thanksgiving dinner for your entire family, only to have the turkey and stuffing go up in flames 2 hours before everyone rang the doorbell, and every store in town is closed. I was devastated and did not know what to do. The data ended up taking another five months to re-create and in the meantime, certain items needed to be tweaked. The social media integration was not working correctly when

I downloaded the test app to my phone, so that took a few emails back and forth over several days to work out. Then, the characters of Bus Rage did not match up on a few levels, so those bugs had to be worked out. It seemed that once one problem was resolved, another would pop up. It was a process of re-design, modification of levels, and compromise of many, many items. It was like each batch of oat meal cookies burned, time after time and even though I watched them closely, the bottoms continued to come out black and burned. Then, one day they were perfect, and I saw the light at the end of the tunnel. That day, I tested the app, went all the way through the levels, had my notepad by my side to write down bugs and discrepancies, and there were none. None! I played it again. I deleted it from my phone, downloaded the latest test version again and voila! It worked again. Much to my surprise it was finally, after months and months of delays, ready to be sent to the App Store.

In order to test your app, the app developer company will send you the app to load onto your smart phone to test, and re-test. This is a fairly simple process and involves downloading it on to your computer, to your desktop, then uploading it to iTunes Store under the App tab, then synching it to your phone and waiting it for it to download. You will have it on your phone, but it will not be "live" on the app store at this point. This is just a test version and will only be available on your own phone for now. The download process from the time you receive it from your developer, until the time it is actually on your phone takes about 45 minutes to 1 hour, each time they send you updates to test. The final product can take up to one hour to download, depending on how big your app is. It is the coolest thing to actually *see* your app on your phone. You will be able to test it just as you would if you downloaded it from the App Store. You will see the icon, and it will look exactly like it should look if you had downloaded it directly from the Store.

During the testing phase, make sure everything works, because this is what will go out to the App Store and what everyone in the world will potentially download. Go through the entire app three times, at least, and write down any tweaks that need to be taken care of. Send a list of fixes to the developer and they will re-send you the version with the revisions.

Every time I downloaded the test version, and I found bugs and sent it back to the development team, I would delete that version of my phone and iTunes so that I was not confused with many different versions. Sometimes you will get 5-10 versions in a week, downloading the test app once or twice a day, and you need to be keeping very organized and make sure the latest version is the one you are actually testing. I found it easier to "start from scratch" every time by deleting the prior versions, but that is really up to you.

Over time, the app will evolve and when it is perfect, it is time to send it to the App Store.

Before I sent my app to the App Store, however, I needed to open an Apple Developer account, which consists of paying $99 and completing the App Store application process. Since my company, Dryven, Inc., was going to be the App Developer, I needed to fill in my information, as well as my company's information. Apple made phone calls and verified the company, made sure I was the true owner, and researched my state's Division of Corporations to make sure the company was legitimate. To this end, if you are going to create a company to "own" your app, it is best to be sure everything is in place prior to you opening up your Apple Developer Account. Make sure you are registered with your State's Division of Corporations and obtain your company's EIN (Employer Identification Number) from the IRS first. Open up a bank account also, under your company's name, as you will need to fill out the banking information and need to put in your company's bank account number so that Apple can automatically and electronically deposit money into your company's account. I had completed the paperwork for Dryven, Inc. months prior, so I had everything in place. You can open a corporation pretty easily and it only takes a few days, so it is up to you whether to have everything in place first, or when the app is ready, do it at that time.

The actual submission to the App Store is exciting.

I had my development company submit my app to the App Store for me. I did have to give them administrative rights and my user name and password. However, this only gave them rights to certain parts of the account, and not the banking part, so I felt it was safe. Once it is uploaded, iTunes Connect will send you emails with updates on what happens and when. For example, you will receive a notification that says something like "your app has been submitted," "your app is under review," "your app is ready for upload to the app store," "your app has been rejected," and so forth. They are very good about keeping you updated throughout the process. My first submission to the App Store was exciting and I almost immediately received an email stating it was received. It took about two weeks for the email to come through that it was under review. Then, the devastating email that said it was rejected. Bus Rage was rejected! That never, ever entered my mind that it would be rejected. Of all the strange and useless apps out there, Bus Rage served a purpose (in my eyes)! It got rid of rage in a safe, fun and entertaining way. I did have options, thank goodness, and the App Store had instructions for their appeal process. I read the instructions over and over, to make sure I fully understood how to appeal, and set off to do so. I wrote for two evenings and stated my case of why Bus Rage was wonderful, why it was so engaging, and how it was a different game each and every time. I submitted the appeal and waited. And waited.

For one week it was torture, waiting for that email notification from Apple. Then, it came. I was so excited I started sweating and re-read it over and over about ten times to make sure I was really reading it correctly. It was accepted, being uploaded to the App Store and was ready for sale just one day later. How excited I was! My first app, in the worldwide App Store. Me, a single mother of two teenagers, working full time, on a budget, and not exactly the most "techie" person in the world, had created an app! I was on top of the world. If I can do it, you can too! If you have a great idea, go for it. Never let go of your dream idea for an app! Despite setbacks, and rejections, never give up because it WILL happen to you too!

Obtain your source code at this point. The App is completed and it is for sale on the App Store. It is yours, so ask your developer for the source code. It will come in an email, most likely, with an attached file. Download that file to your computer, and then create a back-up of it, as well.

DESSERT

Now it was time for dessert, the fun part in my opinion! Chocolate cake, vanilla icing with a cherry on top. Perhaps a scoop of ice cream too. The dessert is the favorite and most important part! Sure, the dinner party is grand. The main course is often raved about, but it is the dessert that is the crown jewel of the evening. A wedding is highlighted by the wedding cake. Dinner on a cruise ship is topped off by succulent desserts. A candy bar at a Sweet Sixteen party is often talked about for weeks afterwards. In keeping with this dinner party spirit, in the world of app creation, marketing, social media and obtaining word of mouth for your yummy app is the "dessert" of app creation.

At this point, you need to market, market, market. The sheer number of app users around the world is staggering. If you reach a mere 2% of the app store users, you are an instant millionaire. However, what seems like a sure thing is not always the case. Sure, Angry Birds is one of those lucky apps that simply exploded. Not all apps are like that, and I found out Bus Rage was harder than I initially thought to market and get the word out.

Marketing is really up to you, and you can go small or you can go really, really big, it is up to your risk factor and how much money you are willing and able to spend. There are companies who go full speed ahead with marketing for a short amount of time, and some companies that gradually and methodically get the word out for their app. I am not sure which way is best, but I have had luck in gradually introducing my app across many social media sites, blog comments, advertising, and websites over a longer amount of time. Right now, my app has been out four full months and has a steady stream of downloads. Nothing crazy and no surges yet, but definitely steady. The time that it takes to post to sites, post status updates, and so forth, is very time consuming, so this will depend on how much time you have to devote to marketing. I happen to work full time outside of my app business, and I used my time from 6 p.m. until midnight to work on it and then in the mornings before work. It was tough, but the steady downloads are worth it. There are a ton of free places to place your app on the internet, and get the word out for free, so definitely do your homework on this.

I scoured the internet to find different ways to market Bus Rage which I will explain here.

Look at the app review sites and inquire their site administrator to see if they will review your app. Some app review sites review for free, others charge a nominal amount.

Changing your price point triggers new app sites to list price reductions on their sites and lists, and can provide much needed exposure.

Search the internet for free marketing ideas and techniques, and for free app developer advertising. Some sites have nominal costs to advertising, such as $50 for a 3 month banner ad. If the website obtains 50,000 visitors per day, this may be worth it since it is such a nominal price. I did this and it definitely created some downloads I would not have otherwise received.

Social media sites, entrepreneur blogs, app developer forums, app developer social networks, and new business blogs often welcome comments where you can provide your website's link or the link to the App Store.

Also, creating your own website explaining your app, what it does, providing contact information and uploading some images of your app, can provide people browsing the web with enough initial information to create intrigue. A link directly to the App Store from the web site will drive traffic towards the app, and more importantly, downloads.

Business cards are always a tried and true method. I used FedEx Office to purchase 1,000 business cards for relatively cheap. I gave them to everyone! When I paid a bill at a restaurant, I left one in the bill pocket. When I got my car serviced, I tacked one on to the bulletin board. When I was at the local fresh market, I would take a local merchant's business card and ask them to take mine. I gave one to everyone I met. I am normally a shy person, but I am sure I received downloads from this simple act of handing out cards.

Content is important for your website, social media sites, and blog comments. Make sure you are clever, witty, and try to incorporate the content with real life events. This will make it more engaging to the customer. For example, I launched Bus Rage right around the time of March Madness and I encouraged fans to "Bus Rage" the other team's players and coaches. I also told them if you didn't like the obnoxious fan next to you in the opposing team's jersey, that you could "Bus Rage" him and feel instantly better.

AFTER DINNER SPIRIT

Gentleman and ladies alike love their after dinner port or brandy! A Godiva chocolate martini is always a great way to top off the evening as the stars come out as well! This is what I call the "extra." You do not need to serve it, but it sure is nice and guests appreciate it. It will keep guests engaged and encourage them to hang around longer. You definitely want your players and users to hang around longer, be engaged and otherwise, be sorry to leave the app to go to sleep (or work). In that sense, the after dinner spirits of app development is tweaking your versions. Play around with the price of the app, decide whether to offer free versus paid, and redesigning future versions of the app. Decide whether to put in more money for new updates, whether to make a "sister" app. Basically, make a plan going forward to sustain interest in the app.

At first, I placed Bus Rage in the third tier of pricing, so it initially started at $2.99. When thirty days went by and I did not really see an increase in downloads, I dropped it to $1.99 with the intent of keeping it there for another thirty days, to see how it progressed. That also did not work too well, much to my frustration. I decided to drop it to 99 cents to see what that triggered. That seemed to work much better and I had many more downloads after that. I think at last count, individuals in 20 different countries have downloaded Bus Rage. The downloads increased after I dropped it twice, however, and this is something to think about when pricing your app. It is a work in progress and sometimes trial and error.

Apple will pay you directly into your developer account. When you are accepted as a developer, the Apple Developer site will ask you to put in your banking information so that they can direct deposit right into your personal or business account. They usually pay once every quarter and you will receive an email indicating you have been paid, and you will also see the deposit in your account. It is pretty cool to see deposit from "Apple" in your bank account and the first one was pretty exciting.

If your app does very well, you may want to consider a "sister" app. For instance, Bus Rage is going to have a separate version in the App Store in about a year. Just like Angry Birds has the "space" version and other versions, you can have other versions also. This keeps the interest level alive for your players.

Another strategy you can use is to add levels, items or updates to your existing app. I plan to add levels, characters and customizable music to Bus Rage when I recover financially from the initial startup costs. Hopefully this will keep users interested and coming back for more.

Now it is the end of the night and every single guest had a magical night! Looking back, it didn't matter that dinner was half hour late, or that a certain appetizer did not look quite right. You brought it all together, dinner turned out delicious, and the night was a success. You crawl into bed that night with a grand feeling of "I did it!" You sleep blissfully dreaming of your next big holiday or theme party, because now you know that you can do it and have learned so much along the way.

CONCLUSION

In summary, the steps you will go through to create your app are:

- Get the idea
- Protect the idea
- Research the idea
- Create your App Plan and Name your App
- Hire your App Developer
- Stay in tune with the entire creation process with the App Developer
- Create an Apple Developer Account and complete all required contract criteria
- Set up your corporation and banking information
- Test your App
- Finalize your App for submission
- Upload your App
- Wait for Acceptance from Apple
- Obtain the source code
- Celebrate your app in the App Store for millions to see!
- Market
- Create updates, different versions and play with pricing of the app

I hope that this will give you an idea of what steps you will need to take, and what to expect, when creating your app from scratch. It is a great feeling to begin the app process and then see it on the app store, on websites, mentioned in app review blogs, and so forth. It is exciting, frustrating and tests your patience, but in the end, you are the creator of an app for the world to see. You are an entrepreneur!

Now, go create your own dinner party, from beginning to end! Have fun planning it, deciding what your menu will be, pick the theme and ambiance and have the best night of your life!

www.ingramcontent.com/pod-product-compliance
Lightning Source LLC
Chambersburg PA
CBHW071826170526
45167CB00003B/1438